BOB ARTLEY's

Seasons
on the FARM

BOB ARTLEY's

Seasons on the FARM

From the Newspaper Series "Memories of a Former Kid"

By Bob Artley
Foreword by Paul Gruchow

Voyageur Press

Cover Design by JoDee Mittlestadt
Printed in China

03 04 05 06 07   5 4 3 2 1

Library of Congress Cataloging-in-Publication Data available
Artley, Bob.
  [Cartoons]
  Bob Artley's seasons on the farm / by Bob Artley ; foreword by Paul Gruchow.— Reprinted ed.
      p. cm.
  "From the Newspaper Series "Memories of a Former Kid"."
  Originally published by the Iowa State Univ. Press in 1981 as Cartoons.
   ISBN 0-89658-609-X (pbk. : alk.paper)
   1. Artley, Bob. 2. Cartoonists—Iowa—Biography. 3. Farm life—Iowa—Caricatures and cartoons. 4. American wit and humor, Pictorial. I. Title: Seasons on the farm. II. Title.
  NC1429.A76A2 2003
   741.5'973—dc21
                        2003008397

To Ginny

Distributed in Canada by Raincoast Books, 9050 Shaughnessy Street, Vancouver, B.C. V6P 6E5

Published by Voyageur Press, Inc.
123 North Second Street, P.O. Box 338, Stillwater, MN 55082 U.S.A.
651-430-2210, fax 651-430-2211
books@voyageurpress.com
www.voyageurpress.com

Educators, fundraisers, premium and gift buyers, publicists, and marketing managers: Looking for creative products and new sales ideas? Voyageur Press books are available at special discounts when purchased in quantities, and special editions can be created to your specifications. For details contact the marketing department at 800-888-9653.

Contents

Foreword

These drawings speak to us out of the universal province of memory.

Some of what we read strikes us as pretty but artificial; we enjoy it and forget it. Some of what we read strikes us as incredible or bizarre; these things we easily forget, too. Some of what we encounter in books or newspapers seems simply beyond judgment—irrelevant, or unremarkable, or outside our experience—and that we pass by.

But when the fully realized artist speaks to us, he speaks directly to our own memories. When we read such an artist, we remember thinking the same thought once, or experiencing once an identical moment, or having once just such an emotion. We are reminded by these shocks of recognition of what Joseph Conrad called the "solidarity of mankind." We are reminded of the tie that binds all humans together—the dead to the living, and the living to the unborn. This is why art is our great weapon against loneliness, which is the beginning of despair.

This may seem too much baggage to attach to Bob Artley's "cartoons," but it isn't. We may think we are charmed and moved by them because they are nostalgic, because they hark back to a quieter and lovelier time, because they are about Iowa, or about farm boys, or something. It is only incidentally so.

These drawings are, in fact, much better than that. At their core, they are not at all about long-ago farm life. They are about growing up, one of the half-dozen great themes of art, a theme Artley has addressed in a fresh and beautiful way. As his title suggests, he has discovered the child buried within himself, and he has dared to share that child truthfully with the rest of us. He reminds us of our own child-ness. He has put us into joyful touch with that forgotten part of ourselves.

For this, we are in Artley's immense and long-lasting debt.

PAUL GRUCHOW

SPRING

THE FIRST HONKERS
OF THE SEASON

15

18

19

24

25

27

WE NEVER DREAMED THERE'D *EVER* BE SUCH A THING AS A FERTILIZER SHORTAGE!

BOB ARTLEY

28

29

33

LAST DAY OF SCHOOL

39

40

41

43

44

45

ONE OF THE MOMENTOUS OCCASIONS OF SPRING—WHEN WE TURNED THE CATTLE OUT TO PASTURE FROM THE CONFINES OF THE MUDDY BARNYARD

48

SUMMER

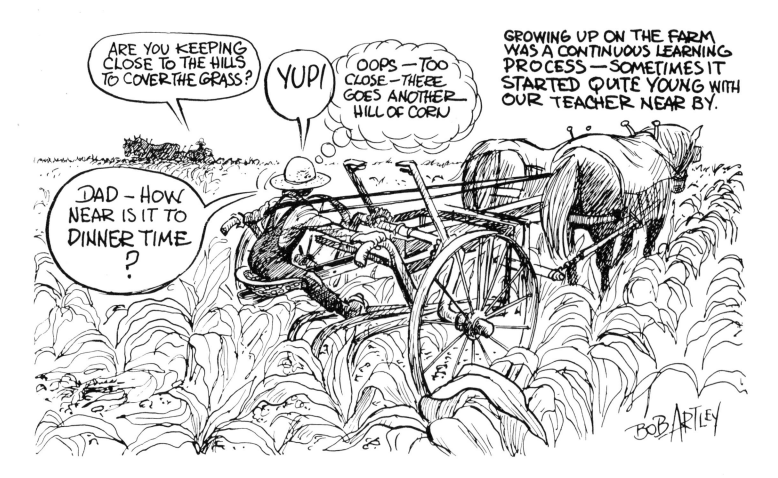

54

ON THE ONE-ROW CULTI-
VATOR I FELT LIKE A
SULKY DRIVER AT
THE COUNTY FAIR AS
I RACED (AND LOST) AGAINST
A SUDDEN SUMMER SHOWER

BOB ARTLEY

WHEN WE'D 'TAKE A SHOWER' AFTER A LONG, HOT DRY SPELL

BOB ARTLEY

60

THERE WERE TIMES WHEN THE SIMPLE TASK OF GOING ACROSS THE SLOUGH AFTER THE MILK COWS COULD BE COMPLICATED AND EXCITING

63

64

66

75

76

77

78

80

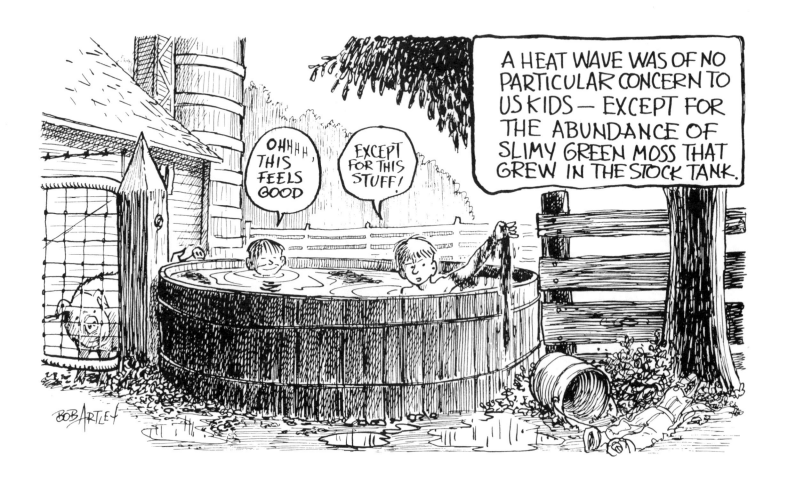

88

93

95

96

98

USUALLY, AFTER THRESHING TIME AND BEFORE SCHOOL STARTED IN THE FALL, WE'D TAKE ONE DAY VACATION TRIPS TO MASON CITY, WATERLOO, CLEAR LAKE OR SOME OTHER NEARBY PLACE OF INTEREST — OUR REST STOPS WERE AT COUNTRY SCHOOL HOUSES ALONG THE WAY

BOB ARTLEY

100

101

AUTUMN

105

107

111

114

115

116

118

121

126

127

129

131

132

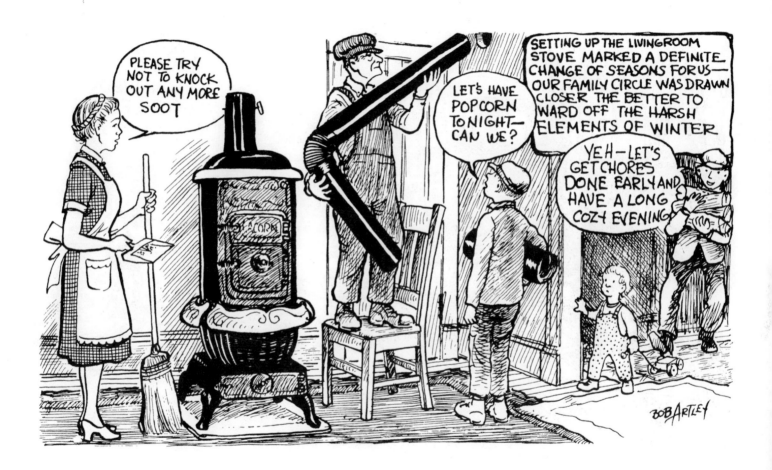

135

OUR AUTUMN HARVEST WAS BY NO MEANS CONFINED TO THE CORN FIELDS . . .

136

140

141

142

143

WINTER

147

148

150

151

152

153

154

155

156

157

158

159

164

167

WHEN WE WERE RESCUED
BY DAD FROM OUR BLIZZARD-
ISOLATED SCHOOL—
NO RESCUE AT SEA COULD HAVE
BEEN MORE WELCOME

169

170

172

174

OUR WINTER RECREATION WAS SOMETIMES BY MOONLIGHT AND INCLUDED THE ENTIRE FAMILY AS WELL AS OUR PETS — ON A LARGE HOMEMADE PONY SLED

179

181

182

183

THE GRILL TOP OF THE NEW SCHOOL STOVE WAS GREAT FOR DRYING OUT OUR WET KNIT MITTENS — BUT WILBUR'S NEW LEATHER MITTENS COULDN'T TAKE THE HEAT

186

NONE OF US HAD ICE SKATES, BUT THAT DIDN'T KEEP US FROM HAVING FUN ON THE FROZEN POND THAT SOMETIMES FORMED AT THE BACK OF OUR COUNTRY SCHOOLYARD — IT WAS ONLY THE DARING (AND FOOLISH) WHO TRIED IT WITHOUT OVERSHOES.

BOB ARTLEY

189

190

192